Animals Adaptations

by Kate Boehm Jerome

Table of Contents

Develop Language 2

CHAPTER 1 Classifying Animals 4
  Your Turn: Classify 9
CHAPTER 2 Different Kinds of Adaptations ... 10
  Your Turn: Interpret Data 15
CHAPTER 3 Using Adaptations to Survive 16
  Your Turn: Communicate 19

Career Explorations 20
Use Language to Restate 21
Science Around You 22
Key Words 23
Index 24

DEVELOP LANGUAGE

Animals protect themselves in many ways. When the porcupinefish is in danger, it fills itself with water. Then the fish looks like a spiny balloon, not a good meal!

Discuss the pictures with questions like these.

Compare the uninflated porcupinefish and the inflated porcupinefish.

The uninflated porcupinefish _____ , but the inflated porcupinefish _____ .

What part of the green sea turtle's body protects it?

The green sea turtle's _____ .

How does the octopus protect itself?

Think about other animals you know. How do they protect themselves?

green sea turtle

Animals: Adaptations

CHAPTER 1

Classifying Animals

The egret's traits include feathers and two legs.

The alligator's traits include scales and four legs.

All animals have **traits**, or characteristics. Traits help tell one kind of animal from another. For example, a bird has feathers and two legs. An alligator has scales and four legs.

Scientists use traits to classify animals into a **classification system**. Animals with similar traits are classified together.

traits – characteristics of a living thing
classification system – a system for grouping things

KEY IDEA Animals with similar traits are classified together.

The animal classification system has many levels. The **kingdom** level is the largest and it contains all animals. Each level underneath the kingdom level contains fewer kinds of animals. This is because each smaller level requires that animals share more traits.

The **species** level is the smallest. It contains animals that are most alike. Every kind of animal is in a separate species. For example, only polar bears are in the species level for polar bears.

Every kind of animal has a **scientific name**. This name includes the genus level and species level names of the animal. Look at the chart to see how a red kangaroo is classified and named.

	Animalia
PHYLUM	Chordata
CLASS	Mammalia
ORDER	Marsupialia
FAMILY	Macropodidae
GENUS	*Macropus*
SPECIES	*rufus*

---

kingdom – the largest level in the animal classification system

species – the smallest level in the animal classification system

scientific name – the name for an animal species that includes the Latin genus and Latin species names of the animal

▶ **The scientific name of the red kangaroo is *Macropus rufus*.**

Invertebrates

All animals are in the animal kingdom. But there is one trait that immediately separates animals into two large groups. This is a backbone, or a line of bones running down an animal's back.

Animals that do not have backbones are **invertebrates**. There are millions of invertebrates that live all over Earth. In fact, over 95% of all animals are invertebrates. The chart shows just a few different invertebrate groups that you might know.

invertebrates – animals that do not have backbones

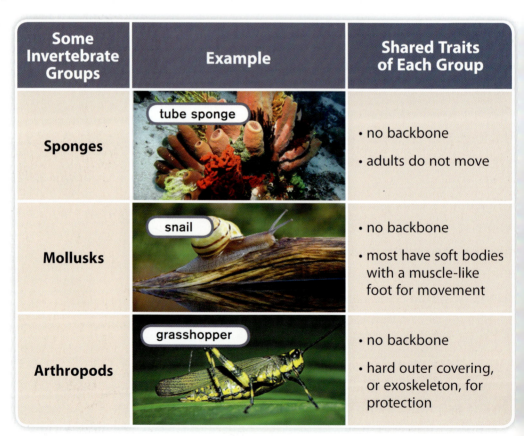

Some Invertebrate Groups	Example	Shared Traits of Each Group
Sponges	tube sponge	• no backbone • adults do not move
Mollusks	snail	• no backbone • most have soft bodies with a muscle-like foot for movement
Arthropods	grasshopper	• no backbone • hard outer covering, or exoskeleton, for protection

6 *Animals: Adaptations*

Ectothermic Vertebrates

Animals that have backbones are **vertebrates**. In most vertebrates, the backbone is part of an endoskeleton, or a larger system of bones. The endoskeleton is inside the animal. It grows as the animal grows.

Some vertebrates and all invertebrates are **ectothermic**. Their body temperatures change with the outside temperature. For example, if the outside temperature is cool, a lizard's body temperature will be cool. If the lizard lies on a warm rock, its body temperature will get warmer.

▲ A frog's endoskeleton supports its body.

vertebrates – animals that have backbones

ectothermic – having a body temperature that changes with the temperature of the surroundings

Ectothermic Vertebrate Groups	Example	Shared Traits of Each Group
Fish	angelfish	• backbone • gills and fins
Amphibians	frog	• backbone • need water for part of their life cycle
Reptiles	lizard	• backbone • skin covered with scales

Chapter 1: Classifying Animals

Endothermic Vertebrates

Other groups of vertebrates are **endothermic**. These vertebrates keep their body temperature steady by using energy from the food they eat. For this reason, endothermic vertebrates usually have to eat a lot more than ectothermic vertebrates.

> **endothermic** – having a body temperature that is kept steady by using energy from food

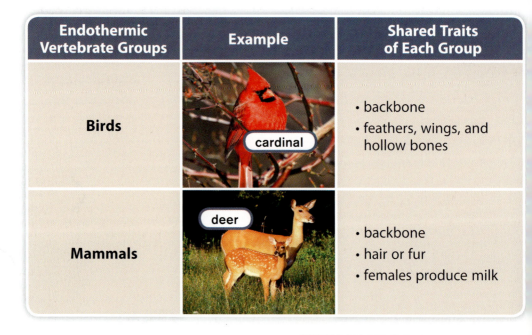

Endothermic Vertebrate Groups	Example	Shared Traits of Each Group
Birds	cardinal	• backbone • feathers, wings, and hollow bones
Mammals	deer	• backbone • hair or fur • females produce milk

Explore Language

The prefix *ecto-* means "outside."
The prefix *endo-* means "inside."

KEY IDEAS Animals are either invertebrates or vertebrates. All invertebrates are ectothermic. Vertebrates can be either ectothermic or endothermic.

YOUR TURN

CLASSIFY

Choose five animals pictured in this book. Classify each animal in a chart like this one. Then talk about your chart with a friend.

Animal	Vertebrate or Invertebrate	Group	Group Traits
snail	invertebrate	mollusks	muscle-like foot for movement; soft body

MAKE CONNECTIONS

When people say "you eat like a bird" they usually mean you don't eat much. Tell why this saying is not really true from a science point of view.

USE THE LANGUAGE OF SCIENCE

What do scientists use to classify animals into groups?

Scientists use similar traits to classify animals into groups.

Chapter 1: Classifying Animals 9

CHAPTER 2

Different Kinds of Adaptations

pit viper

Look at the snake in the photo. The snake's skin color is an **inherited trait**. This means the snake was born with this trait.

The snake's skin color helps the snake hide in trees and bushes. If the snake is hard to see, it can remain safe from animals that might harm it. The snake also has a better chance of surprising the animals that it wants to catch and eat.

Inherited traits that help animals survive are called **adaptations**. The skin color of the green snake is an adaptation.

inherited trait – a trait that an animal is born with

adaptations – inherited traits that help animals and other living things survive

There are different kinds of adaptations. Body parts that help an animal survive are called **structural adaptations**.

For example, the skin color of the green snake is a structural adaptation. The long legs and beak of the flamingo in the photo are also structural adaptations.

structural adaptations – body parts that help an animal survive

The flamingo uses its long legs to wade in shallow water. Its curved beak helps it sift for food in the water.

KEY IDEAS Adaptations are inherited traits that help animals survive. Structural adaptations are inherited body parts that help an animal survive.

Chapter 2: Different Kinds of Adaptations   11

Behavioral Adaptations

The way an animal acts, or behaves, can also be an adaptation. Every animal is born with **instincts**. Instincts, or automatic actions, help the animal know what to do. Inherited instincts that help an animal survive are **behavioral adaptations**.

instincts – automatic actions that an animal is born with

behavioral adaptations – inherited instincts that help an animal survive

SHARE IDEAS Explain how structural and behavioral adaptations are different.

▼ An opossum sometimes "plays dead" when it is in danger. This behavioral instinct may cause its enemies to leave it alone.

KEY IDEA Behavioral adaptations are inherited instincts that help an animal survive.

Learned Behavior

Many people train their dogs not to run in the street. This behavior helps the dog survive. But this kind of behavior is not a behavioral adaptation. The dog was not born knowing how to stay out of the street. The behavior was learned.

Even though **learned behaviors** may help an animal survive, these behaviors are not adaptations. That is because learned behaviors are not inherited instincts.

learned behaviors – animal actions that are not instincts

▲ This dog is learning the hand signal for sit.

By The Way...

Chimpanzees can teach other chimpanzees how to use a tool to get food. But using a stick to catch termites is a learned behavior. It is not a behavioral adaptation.

Chapter 2: Different Kinds of Adaptations   13

Adaptations Work Together

Many times, different kinds of adaptations work together. For example, the armadillo in the photo uses both structural and behavioral adaptations when it is in danger.

The armadillo's body is covered by thick plates that protect it from being eaten. The thick plates are structural adaptations.

But the armadillo also can roll up into a tight ball. This behavioral adaptation works with the structural adaptation to keep the armadillo safe.

Structural and Behavioral Adaptations

▶ A three-banded armadillo rolls into a ball for protection.

KEY IDEA Structural and behavioral adaptations often work together.

14 *Animals: Adaptations*

YOUR TURN

INTERPRET DATA

Walruses, elephant seals, and emperor penguins have adaptations that help them survive in the cold. For example, they can dive deep in the cold ocean to find food. Read the chart to answer the questions.

Animal	Depth of Dive for Food
walrus	80 meters (262 feet)
elephant seal	2,000 meters (6,500 feet)
emperor penguin	500 meters (1,500 feet)

1. Which animal is adapted to dive the deepest for food?

2. Which animal survives *only* on food that is less than 100 meters (328 feet) from the ocean surface?

MAKE CONNECTIONS

When most baby birds hatch from their eggs, they use their beaks to make a hole through the eggshell. Explain why this is a behavioral adaptation.

 STRATEGY FOCUS

Visualize

Look at the photographs in this chapter.
Describe the animal adaptations you can observe in the photos.

Chapter 2: Different Kinds of Adaptations

CHAPTER 3

Using Adaptations to Survive

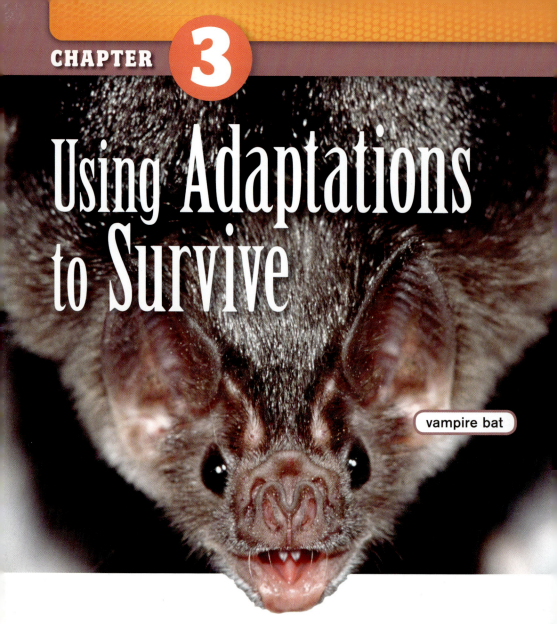

vampire bat

Adaptations help animals do many different things. For example, many adaptations help animals get food.

This vampire bat's nose helps it find food. Its nose can find a warm blood vessel on an animal. Then the vampire bat knows where to use its sharp teeth to make a tiny hole. When blood flows out of the hole, the vampire bat uses its tongue to lick up its meal of blood.

It is important for animals to find food. It is also important for animals to avoid becoming food for other animals. For this reason, many adaptations help animals avoid predators, or animals that might eat them.

Some animals run away from predators. The American pronghorns in the picture can run very fast for a long time. This often allows them to escape from their predators.

▼ The king cobra can raise itself up to 1/3 of its body length.

Other animals try to scare predators away. The king cobra snake can make the skin on its neck widen into a hood. It also raises its body and hisses very loudly.

▼ These American pronghorns can run at speeds up to 100 kilometers per hour (62 miles per hour).

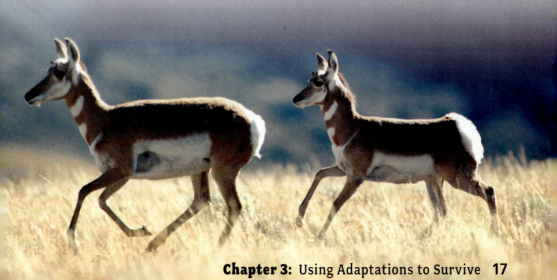

Chapter 3: Using Adaptations to Survive

Camouflage

Some animals hide from predators. Animals often avoid being seen with an adaptation called **camouflage**.

This kind of adaptation allows animals to blend in with the area around them. If camouflage can keep an animal from being seen, the animal has a better chance of survival.

camouflage – special colors, markings, or shapes that allow animals to blend in with their surroundings

▼ Camouflage allows this butterfly to look like a leaf on the tree branch.

KEY IDEAS Many adaptations help animals get food or avoid predators. Camouflage helps an animal blend in with its surroundings.

YOUR TURN

COMMUNICATE

Look at the animals in this book. Choose two animals that move in different ways. Write sentences and draw pictures to explain how each animal moves. Discuss how each animal is adapted to move that way.

MAKE CONNECTIONS

The arctic fox lives where it is cold and snowy in the winter. The fur of this fox is brown in the summer, but the color changes to white in the winter. Give some reasons why this adaptation can help the arctic fox.

EXPAND VOCABULARY

Animals can be **endothermic** or **ectothermic**. The word part *thermo* means "heat." Look up the words shown below. Explain the meaning of each word with words, pictures, or diagrams. Tell how each word is related to heat.

    thermometer   **thermophile**
    thermostat   **therm**

Chapter 3: Using Adaptations to Survive

CAREER EXPLORATIONS

A Marine Park

At a marine park, people take care of sea animals. They help visitors understand more about these animals. There are many different jobs at a marine park. Read the chart. Find out more about these jobs. Tell which job interests you the most.

▲ A trainer works with a dolphin.

Career	Responsibilities
public relations director	• write about the park • appear on TV and radio
dolphin trainer	• train dolphins • take care of the dolphins
videographer	• record videos about the animals

USE LANGUAGE TO RESTATE

Words that Restate

When you restate something, you say the same thing with different words. A restatement can help explain the meaning of a difficult word or phrase. You can use words and phrases such as **or**, **in other words**, and **which is** to make restatements.

EXAMPLE

All animals have traits, **or** characteristics.

Mammals are endothermic. **In other words**, they can keep their body temperature steady.

Look through the book and find places where you can make restatements. With a friend, take turns saying a sentence and restating the information.

Write a Restatement

Choose an animal and find out more about it. Write about this animal's adaptations.

- Describe at least one structural adaptation and one behavioral adaptation of this animal.
- After you describe an adaptation, restate it in different words.
- Draw a picture of the animal that shows its adaptations.

Words You Can Use

or
in other words
which is
which are
this means that

Use Language to Restate

SCIENCE AROUND YOU

You don't have to visit a zoo to see interesting animal adaptations at work. You can observe animal adaptations in a field, a city park, or a backyard. Observe the adaptations in the photos. Answer the questions.

- How does the long beak of a hummingbird help it get food?
- Why do spiders build webs?
- What other adaptations do you see?

Key Words

adaptation (adaptations) an inherited trait that helps animals and other living things survive
An animal **adaptation** often helps an animal get food.

behavioral adaptation (behavioral adaptations) an inherited instinct that helps an animal survive
An opossum uses a **behavioral adaptation** when it plays dead.

camouflage special colors, markings, or shapes that allow animals to blend in with their surroundings
The skin color of a green tree snake is good **camouflage** among the trees.

classification system (classification systems) a system for grouping things
Scientists group animals according to similar characteristics in the animal **classification system**.

inherited trait (inherited traits) a trait that an animal is born with
Skin color is an **inherited trait**.

instinct (instincts) an automatic action that an animal is born with
Eating is an **instinct**.

invertebrate (invertebrates) an animal that does not have a backbone
Over 95% of the animals on Earth are **invertebrates**.

kingdom (kingdoms) the largest level in the animal classification system
All animals are in the animal **kingdom**.

learned behavior (learned behaviors) an animal action that is not an instinct
When a dog is taught a new trick, the trick is a **learned behavior**.

species the smallest level in the animal classification system
Only animals with very similar traits can be in the same **species**.

structural adaptation (structural adaptations) a body part that helps an animal survive
The long legs of a flamingo are a **structural adaptation**.

trait (traits) a characteristic of a living thing
A **trait**, such as number of legs, can help identify an animal.

vertebrate (vertebrates) an animal that has a backbone
A dog is a **vertebrate**.

Key Words   23

Index

adaptation 10–14, 15, 16–18, 19, 21, 22
arthropod 6
backbone 6–7
behavioral adaptation 12–14, 15
camouflage 18, 19
classification system 4–5, 19
ectothermic 7–8, 19
endoskeleton 7
endothermic 8, 9, 19
exoskeleton 6
inherited trait 10, 12
instinct 12-13
invertebrate 6–7, 9
kingdom 5–6
learned behavior 13
mammal 8, 21
phylum 5
predator 17–18
scientific name 5
species 5
structural adaptation 11, 14, 15, 19
trait 4–8, 9, 10
vertebrate 7–8, 9

MILLMARK EDUCATION CORPORATION
Ericka Markman, President and CEO; Karen Peratt, VP, Editorial Director; Lisa Bingen, VP, Marketing; Rachel L. Moir, Director, Operations and Production; Shelby Alinsky, Assistant Editor; Mary Ann Mortellaro, Science Editor; Kris Hanneman, Photo Research

PROGRAM AUTHORS
Mary Hawley; Program Author, Instructional Design
Kate Boehm Jerome; Program Author, Science

BOOK DESIGN Steve Curtis Design

CONTENT REVIEWER
Kefyn M. Catley, PhD, Western Carolina University, Cullowhee, NC

PROGRAM ADVISORS
Scott K. Baker, PhD, Pacific Institutes for Research, Eugene, OR
Carla C. Johnson, EdD, University of Toledo, Toledo, OH
Donna Ogle, EdD, National-Louis University, Chicago, IL
Betty Ansin Smallwood, PhD, Center for Applied Linguistics, Washington, DC
Gail Thompson, PhD, Claremont Graduate University, Claremont, CA
Emma Violand-Sánchez, EdD, Arlington Public Schools, Arlington, VA (retired)

TECHNOLOGY
Arleen Nakama, Project Manager
Audio CDs: Heartworks International, Inc.
CD-ROMs: Cannery Agency

PHOTO CREDITS cover ©Christophe Corteau/NaturePL; IFC and 15b ©David Safanda/iStockphoto.com; 1 ©Pal Hermansen/The Image Bank/Getty Images; 2-3 ©Marty Snyderman/Visuals Unlimited; 2 ©PETER SCOONES/Nature Picture Library; 3a ©Jeff Rotman/Nature Picture Library; 3b ©Jez Tryner/Image Quest Marine; 4a ©Paul Brough/Alamy; 4b ©David Osborn/Alamy; 5, 6b, 22b ©Arco Images/Alamy; 6a ©Andre Seale/Image Quest Marine; 6c ©Gerry Ellis/Digital Vision/Getty Images; 7a illustration by Joel and Sharon Harris; 7b and 7c © Gary Bell/Oceanwide Images; 7d ©George McCarthy/Nature Picture Library; 8a ©Tom Vezo/Nature Picture Library; 8b and 21 ©Claudia Adams/Alamy; 9a and 9b Ken Karp for Millmark Education; 10 ©Nick Garbutt/Nature Picture Library; 11 ©Peter Oxford/Nature Picture Library; 12 ©Steve Maslowski/Visuals Unlimited; 13a ©Juniors Bildarchiv/Alamy; 13b ©Steve Bloom Images/Alamy; 14a, 14b, 14c, 14d ©Mark Payne-Gill/Nature Picture Library; 15a ©Armin Rose/Shutterstock; 16 © Barry Mansell/Nature Picture Library; 17a ©Dinodia Images/Alamy; 17b ©Corbis Premium RF/Alamy; 18 © Cheryl Hogue/Visuals Unlimited; 19a © Dave Watts/Nature Picture Library; 19b © Minden Pictures/Getty Images; 19c ©Fletcher & Baylis/Photo Researchers, Inc.; 20 © Jeff Greenberg/PhotoEdit; 22a ©Ablestock/Alamy; 22c ©Oxford Scientific/Photolibrary; 23 ©Stephen Dalton/Photo Researchers, Inc.; 24 ©Getty Images/Gallo Images ROOTS Collection

Copyright ©2008 Millmark Education Corporation

All rights reserved. Reproduction of the whole or any part of the contents without written permission from the publisher is prohibited. Millmark Education and ConceptLinks are registered trademarks of Millmark Education Corporation.

Published by Millmark Education Corporation
7272 Wisconsin Avenue, Suite 300
Bethesda, MD 20814

ISBN-13: 978-1-4334-0142-8

Printed in the USA

10 9 8 7 6 5 4 3 2 1

24 *Animals: Adaptations*